THE ABC ZOO

The ABC ZOO

Beastly Facts from A to Z

OLIVER WEISS

PA PRESS

PRINCETON ARCHITECTURAL PRESS · NEW YORK

Published by
Princeton Architectural Press
A division of Chronicle Books LLC
70 West 36th Street, New York, NY 10018
papress.com

© 2025 Oliver Weiss
All rights reserved.
Printed and bound in China
28 27 26 25 4 3 2 1 First edition

ISBN: 978-1-7972-3720-6
LCCN: 2024057862

No part of this book may be used or reproduced
in any manner without written permission from the publisher,
except in the context of reviews.

Every reasonable attempt has been made to identify
owners of copyright. Errors or omissions will be corrected
in subsequent editions.

Editor: Allison Serrell
Designer: Oliver Weiss

For Pauline & Nicolas

A

The **ANACONDA** is one of the biggest and heaviest snakes in the world. It silently glides through dense marshes, thick swamps, and deep lakes, hunting under the cover of night.

B

In just two short weeks, a caterpillar undergoes the magical transformation of metamorphosis and begins a new life. **BUTTERFLIES** have delicate wings and come in many sizes. And their wingspans can be quite majestic.

C

The **CROCODILE** stretches out by the riverbank, soaking up the warm sun. He dreams of tasty fish swimming in the water and of colorful birds flying high in the sky.

E

ELEPHANTS are the largest animals living on land. They use their long trunks to communicate with each other and gobble up large chunks of food in big bites.

FROGS start as eggs that hatch into tadpoles. As they grow, they develop legs and lungs and change into frogs. Then they're all about tongue flicking, hopping, and croaking all day.

F

G

With a tongue the length of your arm, the **GIRAFFE** stretches her long, spotted neck high into the sky, snacking on leaves right off the highest branches.

H

HAWKS are birds of prey with sharp vision, curved beaks, strong talons, and long, powerful wings. They can spot a tiny mouse from a mile away!

I

Tiny drops of salty spray are tickling the **IGUANA**'s scaly face as ocean waves crash against the warm, sun-baked rock. She uses her long tail to dive, gliding through the water like a submarine.

JAGUARS are an extremely rare sight in the wild. Their coats are covered with spots. These big cats love to swim, and they eat almost anything.

K

KANGAROOS live in Australia and have strong hind legs that let them leap like athletes. Female kangaroos carry their young in a pouch until they are big enough to hop around on their own.

L

Shaking his mane and letting out an occasional roar, the **LION** prefers to rest during the day and hunt at night. He can eat a lot of meat in one meal.

The **MOOSE** gracefully wanders through marshlands, grazing on leaves and grass. Even though he blends into the forest with ease, his massive size and swaying antlers make him easy to spot.

N

The **NIGHTINGALE** sings a beautiful tune in the garden at night while fireflies gather under the pale moon. It's tricky to catch a glimpse of this shy bird, as he's a master of hiding.

Trying to race an **OSTRICH** would be like racing a train. Their long legs can carry them faster than anyone can run. It's a good thing, too, because ostriches can't fly!

PENGUINS live in Antarctica, an icy place without flowers or trees, where it's very cold. They wear black and white feathers and spend their days diving in the water to catch fish to eat.

Q

The California **QUAIL** wears a head feather that looks like a fancy hat. Her nest can hold many eggs at once, room for a whole classroom of baby quails!

R

RABBITS love hanging out with their bunny buddies. Their ears can swivel to hear sounds from all directions. They nibble on grass all day to keep their ever-growing teeth from getting too long.

S

A **SQUID** changes color and glows in the dark. When in danger, it disappears by releasing clouds of black ink. Unlike octopuses, which have eight arms, squids have eight arms and two tentacles, and their heads are more triangular.

U

Sea **URCHINS** look a bit like hedgehogs, proudly showing off their spines. Living deep in the ocean, they move slowly along the sandy floor and are often found among their many friends.

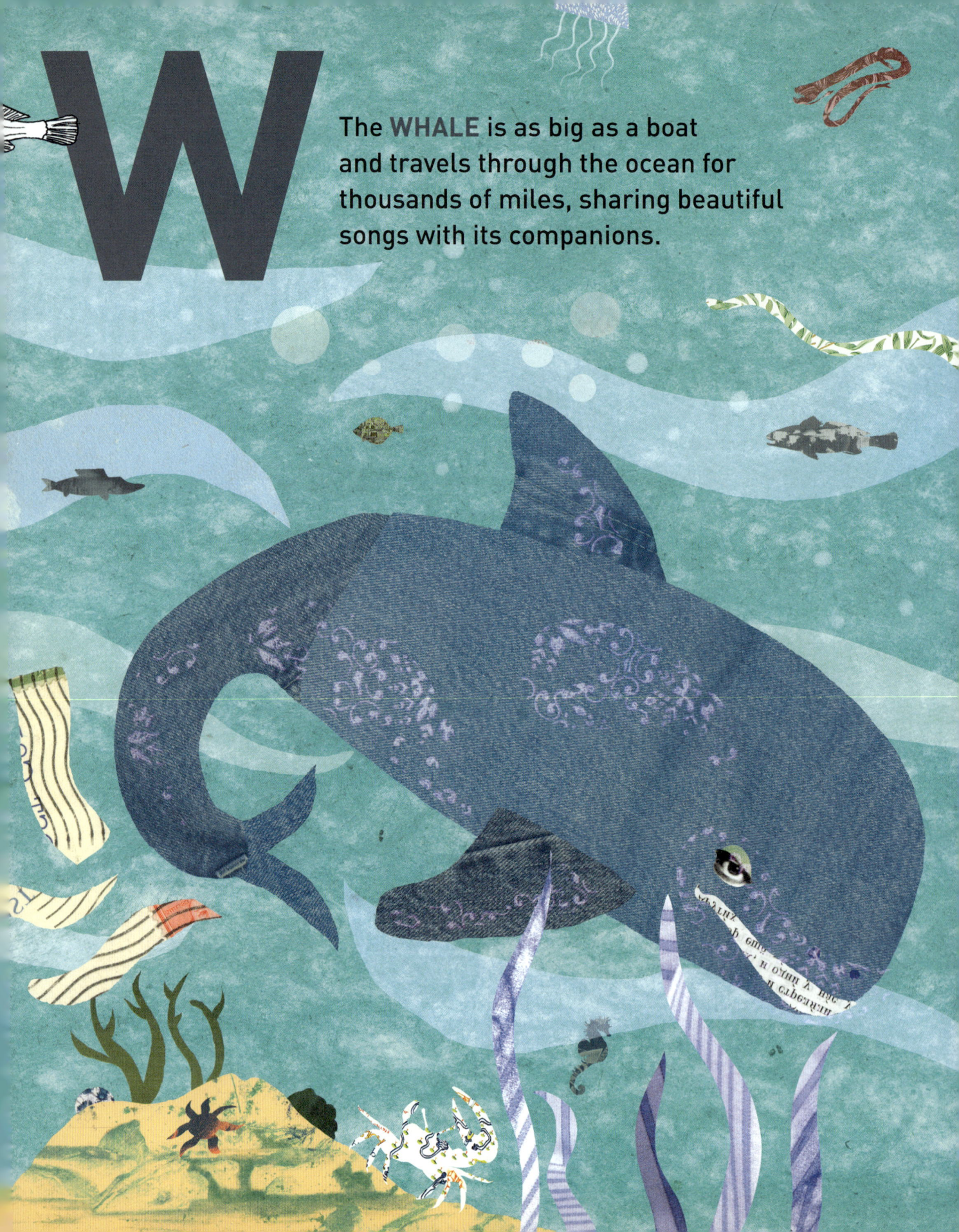

W

The **WHALE** is as big as a boat and travels through the ocean for thousands of miles, sharing beautiful songs with its companions.

X

The tiny **XANTUS'S HUMMINGBIRD** hovers like a helicopter to sip nectar from blossoms, flapping its wings as fast as some insects.

Y

The **YORKSHIRE TERRIER**, one of the smallest dog breeds, loves having its long coat brushed regularly.

Z

ZEBRAS might be mistaken for horses if not for their black and white stripes. These stripes help them hide in the tall grass of the savanna and also confuse flies.

OLIVER WEISS *is an award-winning artist and illustrator who loves to draw and make collages. Ever since he was a child, he has been fascinated by the wild animals around his parents' countryside home, including frogs, hedgehogs, grasshoppers, moles, buzzards, herons, and weasels. He lives in Berlin and New York City.*